Gordon, Jo.

Recycling.

DATE			

RECYCLING

First published in
the United States in 1992 by
Gloucester Press
95 Madison Avenue
New York, NY 10016

Library of Congress Cataloging-in-Publication Data

Gordon, Jo W.
Recycling / by Jo W. Gordon.
p. cm. — (Save our earth)
Includes index.
Summary: Discusses advantages and current methods of recycling and examines possible innovations for the future.
ISBN 0-531-17332-1
1. Recycling (Waste, etc.)—Juvenile literature.
[1. Recycling (Waste)] I. Title. II. Series.
TD794.5.G67 1992
363.72'82—dc20 92-9788 CIP AC

Printed in Belgium

The publishers would like to acknowledge that the photographs reproduced within this book have been posed by models or have been obtained from photographic agencies.

Design	David West Children's Book Design
Designer	Steve Woosnam-Savage
Editors	Elise Bradbury, Michael Flaherty
Picture research	Emma Krikler
Illustrator	Ian Moores

The author, Jo Gordon is the team leader of Waste Watch, the national agency for promoting recycling in Britain.

The consultants: Jacky Karas is a Senior Research Associate at the School of Envrionmental Sciences at the University of East Anglia, Great Britain.

SAVE OUR EARTH

RECYCLING

JO GORDON

GLOUCESTER PRESS

London · New York · Toronto · Sydney

CONTENTS

WHAT IS RECYCLING?
6

SAVING RESOURCES
8

HELPING THE ENVIRONMENT
10

THE PROBLEM OF WASTE
12

RECYCLING AT HOME
14

HOW IT WORKS
16

COMPOSTING
18

INDUSTRIAL WASTE
20

RECYCLING ENERGY
22

RECYCLING TODAY
24

WHAT YOU CAN DO
26

FACT FILES
28

GLOSSARY
31

INDEX
32

INTRODUCTION

Helping to reduce the problems that are affecting our environment is something we all want to do, but it can often be difficult to know where to start. Collecting materials like glass, cans, paper, clothes, and plastics is something we can all do, everyday, in our own homes. When these materials are recycled – that is used again to make the same thing, or made into something new – we know we are saving energy, reducing pollution, helping to conserve valuable natural resources – many of which are in short supply – and reducing the amount of waste that has to be thrown away.

In industrial societies, consumer goods are manufactured at a high rate. People buy new, improved cars, refrigerators or cookers while the old ones are just thrown away. These discarded items contain a lot of valuable materials that can be reclaimed.

Industry produces millions of tons of waste every year including paper, ash, metals, and plastic. Although industry does now recycle to some degree, this could be done on a much larger scale.

Throughout the world, governments are recognizing that recycling is a sensible way to make use of resources that would otherwise just end up on the trash heap. In many countries there are now large scale recycling programs. As it becomes more difficult to find ways to get rid of the garbage we produce, it makes sense to recycle and reuse what we can.

◀ Up to 60% of the waste we produce from our homes and backyards could be reused or recycled into new things. As shown by these people from Bangladesh sorting paper, it is commonplace in developing countries to recycle as much as possible.

WHAT IS RECYCLING?

Recycling is using things again either for the same purpose, or by making them into something new. One of the best ways to recycle glass for instance is to use refillable bottles, like milk bottles – which can be reused several times. Glass can also be melted down and made into new bottles and jars.

Recycling is not a new idea. In nature, many things are recycled over and over again. In the past, things were made to last and people repaired and reused them, rather than throwing them away. Cars, washing machines, furniture, refrigerators and other items can all be recycled. The steel used to make a Rolls Royce might have come from an old washing machine.

In some countries like the United States and some cities like London, compulsory recycling programs have been introduced to make sure recyclables are not thrown away.

▼ In the natural world, elements are recycled with no waste. While animals breathe out carbon dioxide during the day, plants breathe it in. When plants and animals decompose, they give nutrients to the soil which are taken up by plants. Some of the carbon in the remains forms, over millions of years, the basis of coal, oil and gas, which when burned, release carbon dioxide into the atmosphere.

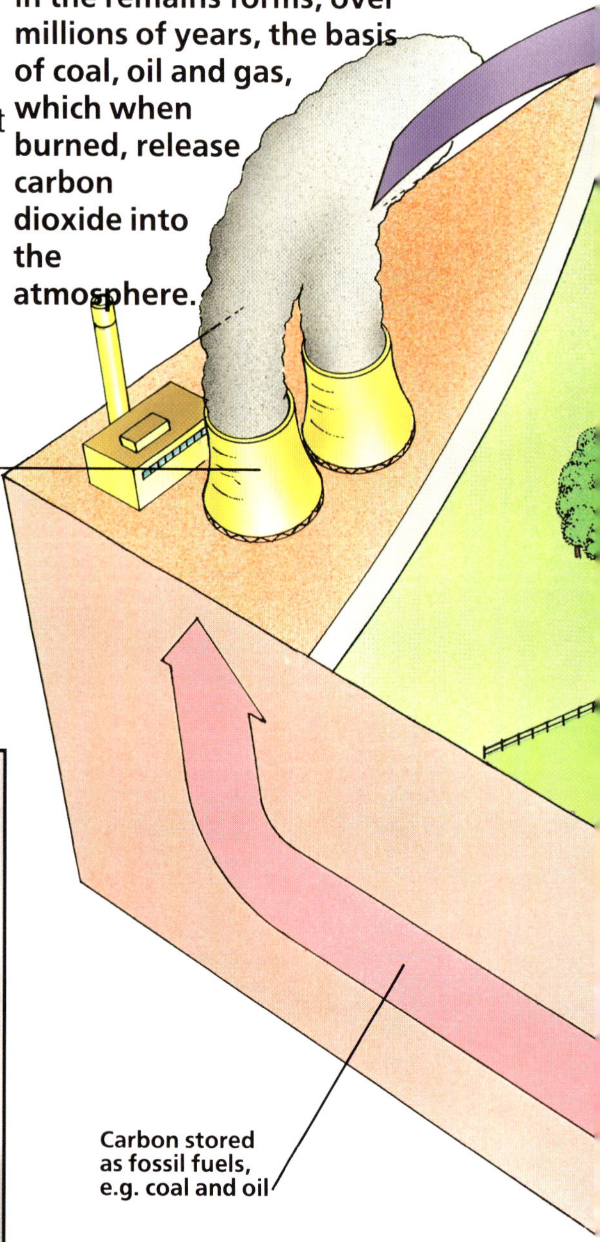

▼ Phoenician boats from the time of the Ancient Greeks have been found laden with glass cullet (broken bits of glass) collected to make new glass objects.

► Recycling is not new. This picture shows pots being melted down for repairs.

THE CARBON CYCLE

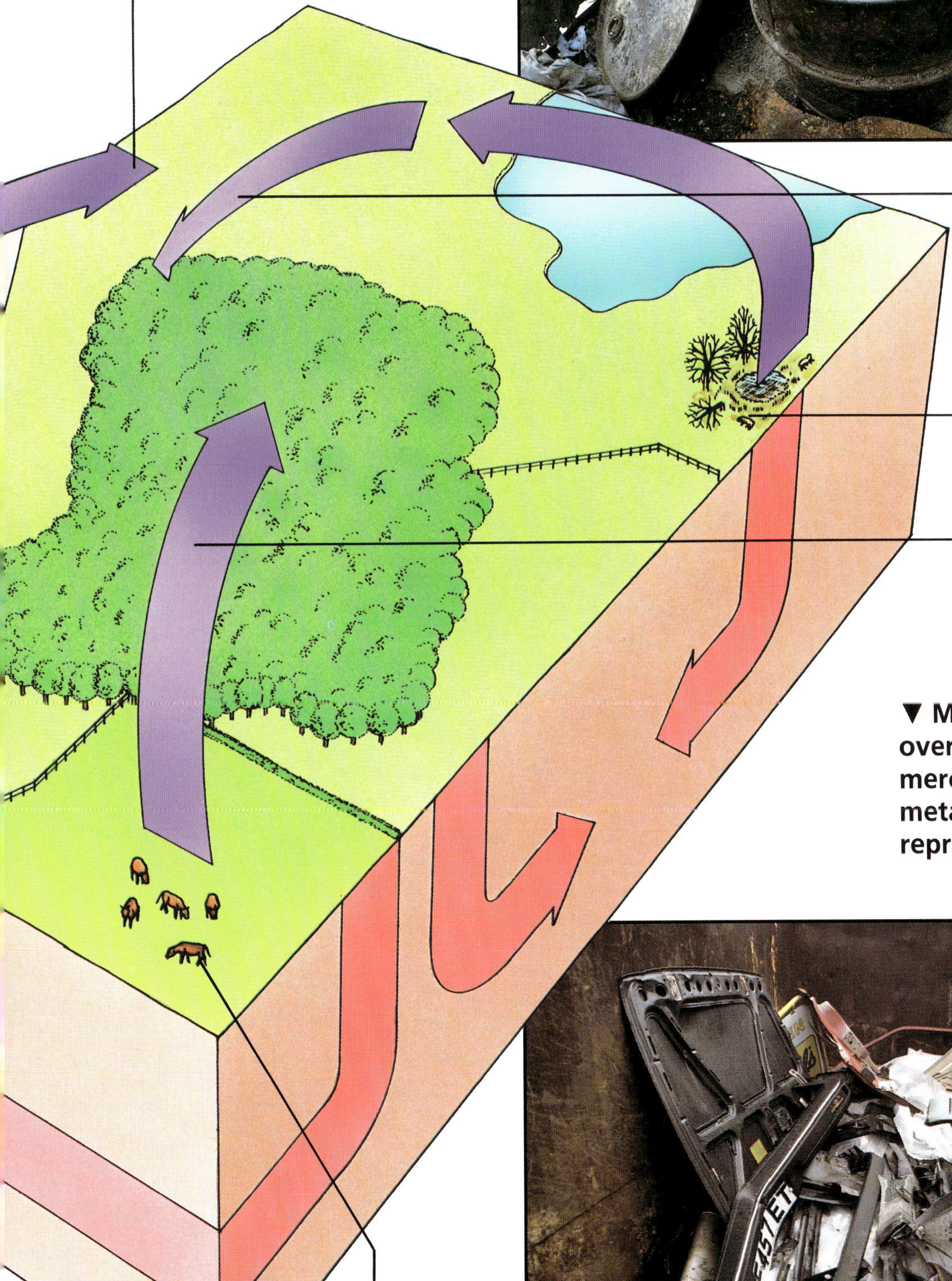

▼ Metals can be recycled over and over again. Scrap merchants collect and sort metals so they can be reprocessed.

SAVING RESOURCES

Making goods from recycled materials can help save fuel and means that valuable resources are not being used up unnecessarily. Many of the things we use in our homes are made from materials like oil and gas which are finite resources. This means there is only a limited amount of them and eventually we will run out of them. Plastics, nylon, fertilizers, and chemicals are just some of the products made from oil and gas.

Making goods from recycled materials can also help to reduce the amount of energy needed to produce them. Energy for manufacturing and transporting goods comes mainly from coal, oil and gas. Making new aluminum cans from old aluminum cans can result in an energy saving of up to 95%. Every ton of glass made from recycled glass cullet saves 30 gallons of oil. Making paper from recycled paper needs only half the energy needed to make paper from wood pulp.

▼ Offshore oil rigs will no longer be necessary, when oil reserves run out. If we continue to use oil at the present rate, it will run out in another thirty years. Recycling oil and gas products will help slow down this drain and save energy.

This china clay pit in Trethosa, and the oil drilling installation in Ecuador (inset), show the scars of mineral and resource excavation. Mining metal ores and other useful minerals also causes terrible damage to habitats, disrupting plant and animal life, and poisoning soil, water supplies, and the air. Recycling these resources reduces the need for such extensive mining, allowing supplies to last longer. This also reduces the destruction to vital habitats, containing precious flora and fauna.

HELPING THE ENVIRONMENT

Recycling can help to reduce both pollution and waste. Less waste to be disposed of means fewer landfill sites or less need for incineration. Less demand for raw materials means that fewer habitats will be destroyed by mining, and valuable fuel and resources will be saved.

Making goods from recycled materials can reduce air and water pollution. It is estimated that using waste paper in paper production, for instance, can reduce water pollution by up to 35% and air pollution by up to 74%. This is important as paper making can lead to the release of sulfur dioxide, which is one of the main causes of acid rain.

Making paper from recycled paper will not save trees – trees are specifically grown to make new paper – but it can stop conifers being grown in areas where they cause damage, such as the Scottish Highlands. Conifers acidify the soil, and herbicides and pesticides used to protect the crop can damage other plants.

► Fast growing trees, such as these eucalyptus trees in Argentina, are being grown for paper making. The trees are not bad in themselves, but because they are being grown in places in which they would not normally be found, they upset the local ecological balance. They acidify the soil, destroying local plants and making it difficult to grow other crops.

▲ Bauxite, the ore from which aluminum is made, is mined in tropical rain-forest areas, an environment that contains many resources, and is itself under threat. This bauxite lake in Jamaica shows the pollution that can result.

▲ Over 15 tons of carbon dioxide are released when 10 tons of iron is made, with 5 tons of coke burned. Collecting and recycling scrap metal is now an enormous industry in its own right. Not only does it require less energy to remake the steel, but it reduces air pollution and rids the environment of rusting scrap.

THE PROBLEM OF WASTE

The amount of waste being produced from homes, particularly in developed countries, is increasing every year. At the same time it is becoming more difficult to find ways of disposing of this waste.

Most waste is disposed of in landfill sites, some is burned in incinerators and a relatively small amount is reused or recycled. The amount of waste produced is also growing because an ever growing range of disposable items are being used, such as plastic cups, cutlery, containers, tissues, disposable diapers, razors, and so on. People in these "throw-away societies" also throw out goods, not because they are worn out, but because they are no longer fashionable. Products are being made so that they actually have to be replaced regularly.

Many of the things in our garbage dumps are valuable and we may find that dumps are mined in the future to retrieve these materials.

► Landfill sites, large holes in the ground, are where most of our garbage will end up. As sites fill up, it is getting harder to find new ones. No one wants to live near a garbage dump and there can be problems of pollution and from methane gas.

▲ Garbage in streets is unsightly and expensive to clean up.

◄ Most solid waste is put into landfills. These can attract birds, rats, and other scavengers causing a health hazard.

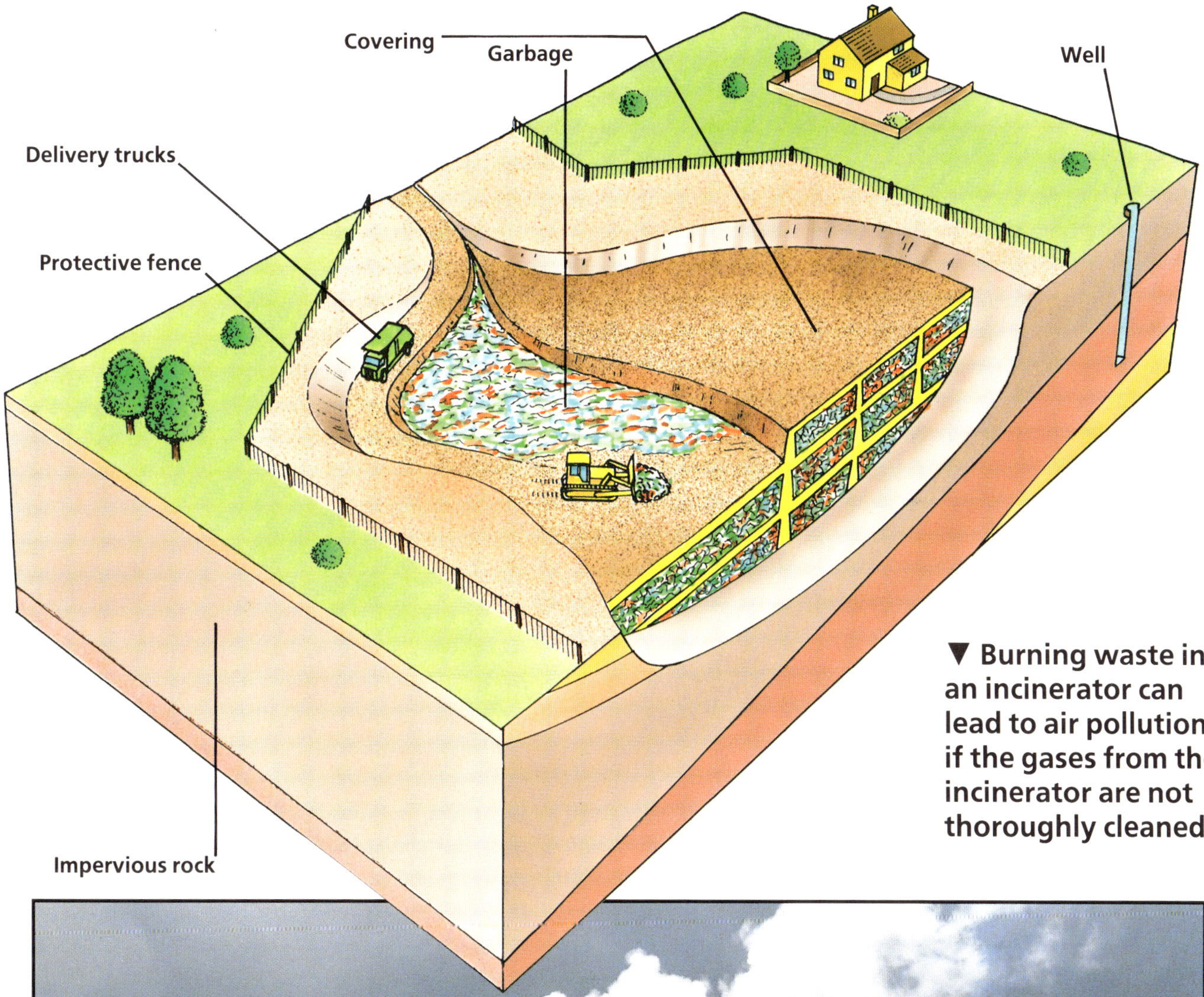

▼ Burning waste in an incinerator can lead to air pollution if the gases from the incinerator are not thoroughly cleaned.

RECYCLING AT HOME

About 80 percent of the trash in our garbage cans could in theory be recycled, creating a significant saving in raw materials and the energy needed to process them. But due to contamination of the rubbish only about 60 percent is clean enough to use. That means we would be able to halve the amount of waste we produce by recycling.

However, what we are able to recycle will depend on the collection facilities available locally. Bottle banks are probably the best known and most widely available collection system worldwide.

In some parts of the United Kingdom, in the United States, Canada and Germany there are "curbside collection" programs where residents put recyclable garbage into a separate box or trash can ready for collecting. In other areas, people save their recyclables and then take them to a recycling center which has different containers for different items.

▼ If you are collecting materials to take to a recycling center, you should try to make sure you combine this with a trip you would make anyway, to the supermarket or school, for instance. Otherwise, if you travel by car you might use more energy and create more pollution than you save through recycling.

Every household in the United Kingdom throws away about a ton of garbage a year. That is the equivalent of two trees worth of paper, 210lb of food and beverage cans, $30 worth of clothing, 250 bottles and jars, and 100lb of plastics. Paper, cardboard, glass, plastic, textiles, cans, foil, and organic waste can all be reused or recycled, thereby reducing the amount of garbage dumped in landfill sites.

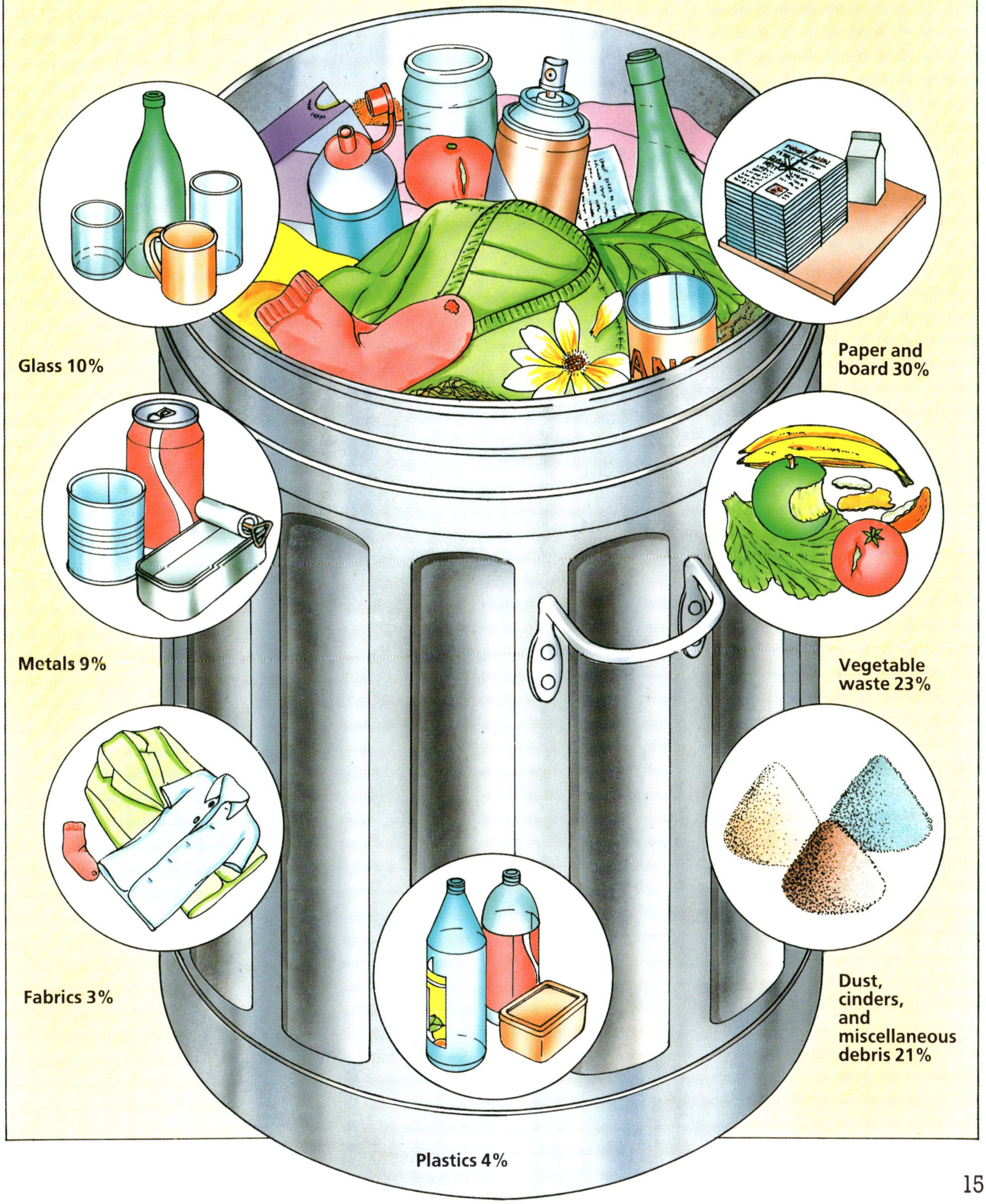

HOW IT WORKS

Once recyclable materials have been collected from homes or recycling centers, they are either transported directly to the reprocessing plant if they are already sorted, or they are taken for sorting, before being reprocessed. The material is then cleaned and any contaminants are removed. Glass is sorted by color – usually green, brown, and clear – and is kept separately to avoid color contamination, before being melted down and made into new bottles and jars. Aluminum cans are crushed and packed into bales which are then melted down to make new drink cans. Tin-plated steel cans are treated to remove the thin layer of tin, and both the steel and tin can be used again.

Newspapers and magazines can be repulped and used to make new newspapers and magazines, or other products such as cardboard. Quality paper can be used to make high grade recycled paper for printing and writing. Plastics can be used as either mixed plastic or sorted into different varieties and then made into new goods.

Only about a quarter of all the paper and cardboard made each year all over the world is made from waste paper. Recycling paper helps reduce waste disposal problems and can reduce water pollution and fuel use when compared with making paper from trees. There is plenty of opportunity for more paper recycling, but it will only happen if the paper industry invests more in recycling plants.

Air classifiers remove light materials first. Heavier items like heavy plastic, glass, and cardboard are separated when they fall through an air jet into various collecting trays. Iron and steel objects are removed by magnetic separators.

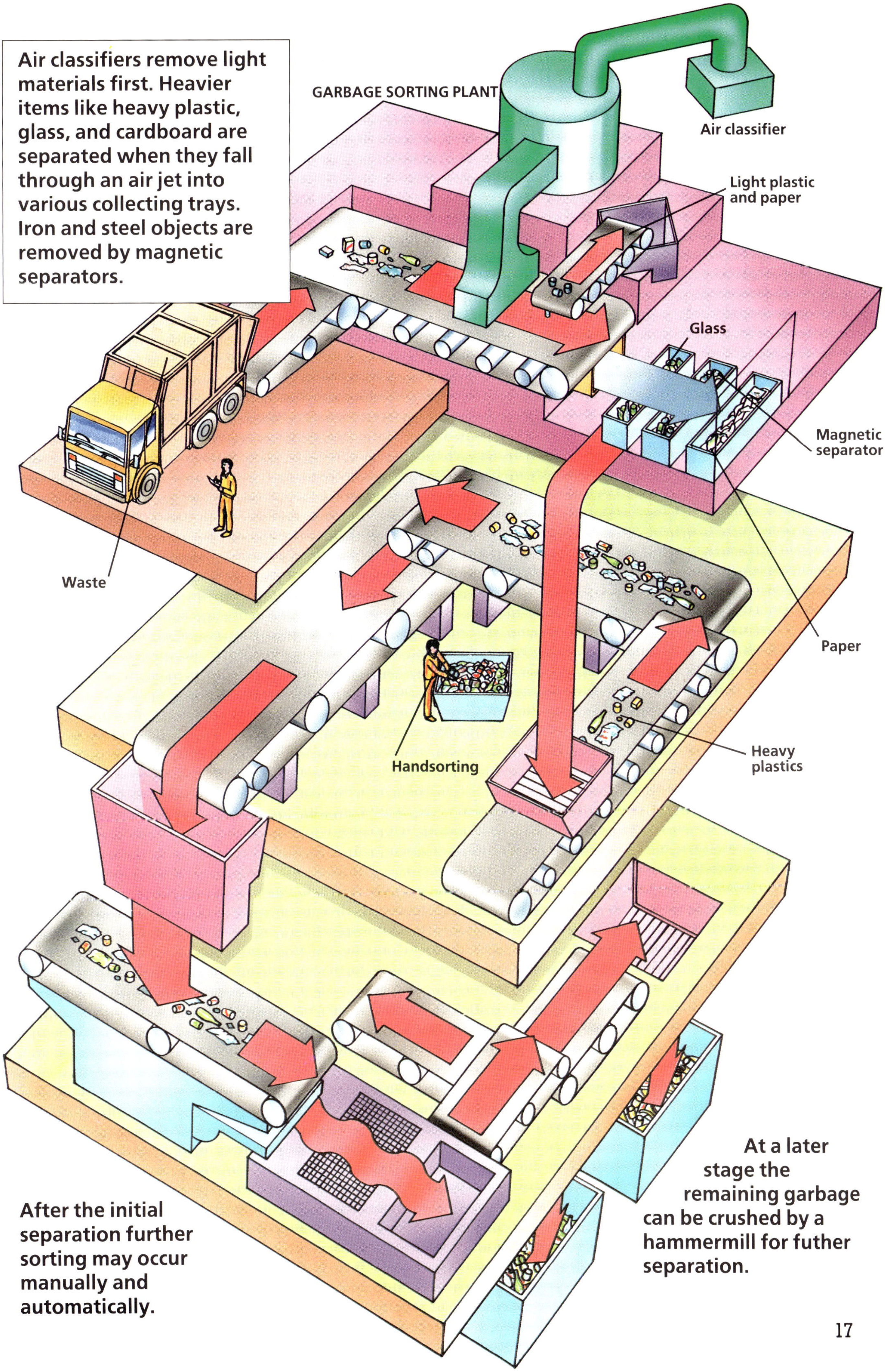

After the initial separation further sorting may occur manually and automatically.

At a later stage the remaining garbage can be crushed by a hammermill for futher separation.

COMPOSTING

Kitchen and garden waste make up about 25 percent of the trash in our garbage cans. This waste, if properly composted, can provide an important source of nutrients for plants, help to improve soil structure, keep moisture in the soil, and protect plants from frost. If you have a backyard or garden it is worth setting up a compost heap. If space is limited, try using an alternative such as a compost pail which needs less room. Composting saves money as it means expensive, potentially harmful fertilizers are no longer needed and it can provide an alternative to peat. Peat-cutting for use in gardens is destroying important and irreplaceable habitats.

In some countries like the United States and Germany, "yard waste" is collected separately and programs to explore composting on a large scale are being developed. Garden and food waste can be turned into the fuel methane gas in biogas generators.

▼ An organic garden is one where no artificial fertilizers or pesticides are used, so no harmful substances enter the soil. Compost can provide all the necessary food for plants and used as a mulch, a covering over the soil which will keep weeds at bay. Garden pests are controlled by using natural methods such as introducing ladybugs which prey on aphids, thus avoiding the need for harmful chemicals.

Composting is not difficult. Compost can be made all the year round, although the microbes in the heap will be most active between April and September. Vegetable peelings, garden waste, and grass cuttings can all be put into the heap.

The compost should be kept moist and turned occasionally to allow oxygen in. The heap will also need to be covered to keep the heat in.

Compost should contain a mixture of materials high in nitrogen and those high in carbon. A good mix helps the microbes that break down the organic material.

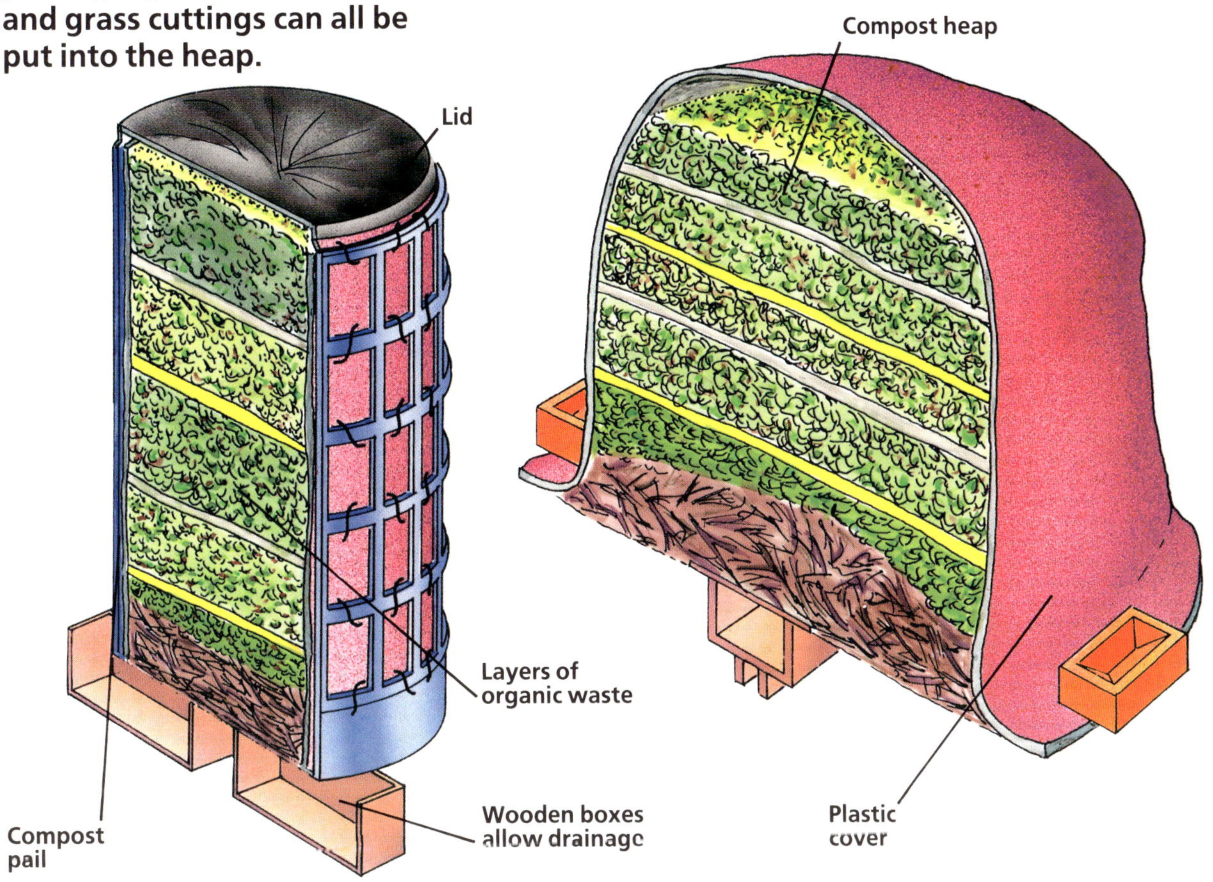

▶ Many farmers in the developing countries, like this one in Indonesia, make good use of compost and animal manure. An increasing number of farmers in the developed world are returning to traditional organic methods, due to concern about the effects of industrial farming methods on the quality of food, soil and water.

INDUSTRIAL WASTE

Industry produces far more waste than people do from their homes. Some industrial waste is potentially dangerous and has to be disposed of carefully. Industry also produces waste gases which, unless properly controlled, can create acid rain and add to the greenhouse effect. The problems of industrial waste can be reduced if industries are encouraged to remove wastes from dirty water and air before releasing them. Careful separation of chemical wastes from waste water can reduce disposal charges and provide useful chemicals for the factory.

In some countries, waste exchange programs have been set up to allow companies to trade waste that would be useful to another company.

In the United Kingdom, Play Resource Centers collect safe, nontoxic waste from firms with the purpose of providing exciting and unusual play materials for children's playgrounds.

▼ Industry is continually looking for new ways to make use of waste materials. An example of this is ash from power stations which can be made into bricks.

Plastics make up a growing percentage of waste as they replace traditional markets. The recyling of plastic thus becomes more important, and special plants like this one have been developed to create new plastic materials like garbage bags and traffic cones, from plastic waste.

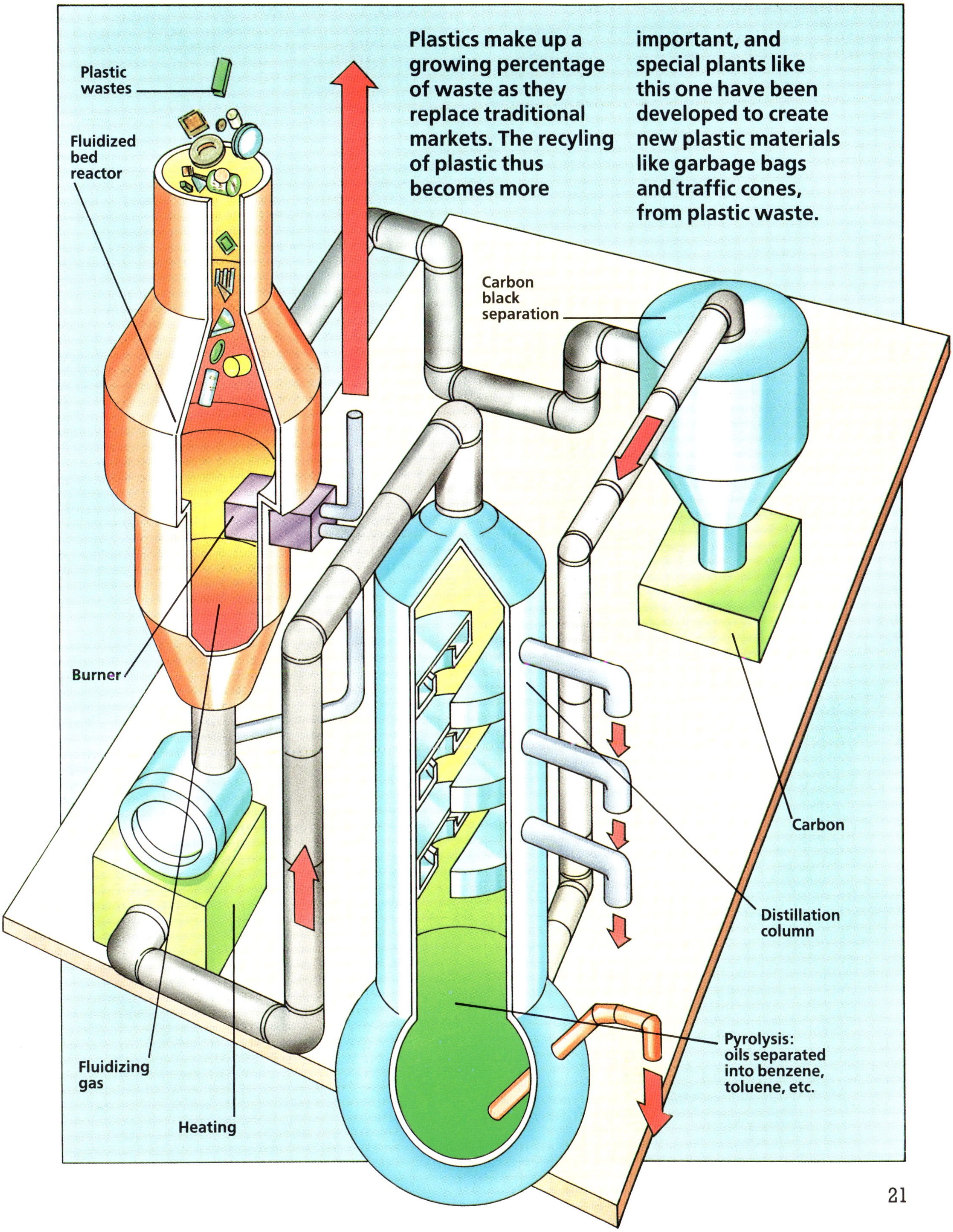

RECYCLING ENERGY

Waste is a potential source of energy. This energy can be recovered when garbage is burned in an incinerator providing that the heat can be used for heating systems or to generate electricity. It is important that materials such as glass and metals, which are nonrenewable resources, are taken out so that they can be recycled. Scientists are looking at ways in which waste can be compressed into pellets which can then be burned in industrial boilers. Experiments have also been carried out to produce oil from kitchen waste by heating it under great pressure.

There is a growing number of resource recovery programs that are designed to burn household garbage to provide heat and power to a local area and recycle what is left over. About 700 such programs are in use worldwide. In Denmark 75 percent of household waste is burned and the heat used to generate electricity. Incinerators are expensive to build and there is a risk that gases from incinerators can carry pollutants such as dioxins which are some of the most toxic synthetically produced chemicals. As dioxins do not readily breakdown in soil or water they contaminate the land.

▼ The control center of this incinerator in France monitors the actual burning process, and controls the level of toxic emissions. Incinerators are being built to generate electricity and heating for local use, although not all incinerators recover energy.

▲ The Puento Hills Landfill Gas Recovery Facility in California generates electrical power from methane gas released from the waste dump.

► This biogas plant, typical of many in developing countries, produces fuel from decaying organic waste. In the Netherlands and Belgium, large biogas plants are used to generate electricity for whole areas.

RECYCLING TODAY

Recycling is important, but it is only one way in which we can help to reduce the amount of waste we produce. It is also important to avoid unnecessary waste in the first place, by thinking about what we buy, and avoiding items that have a lot of unnecessary packaging. Buying milk in refillable bottles (where available) rather than throwaway cartons is an example of this.

Furniture, household goods in reasonable condition, clothing, and other items can be given to charities and other organizations where they will be sold and continue to be useful. Many consumer goods like refrigerators and washing machines can be repaired rather than being placed on the dump.

Buying goods made from recycled materials is also vital as this helps to create markets for recycled products. Buying goods that have been built to last is economical since the items will not have to be replaced so often, and there is a saving on resources.

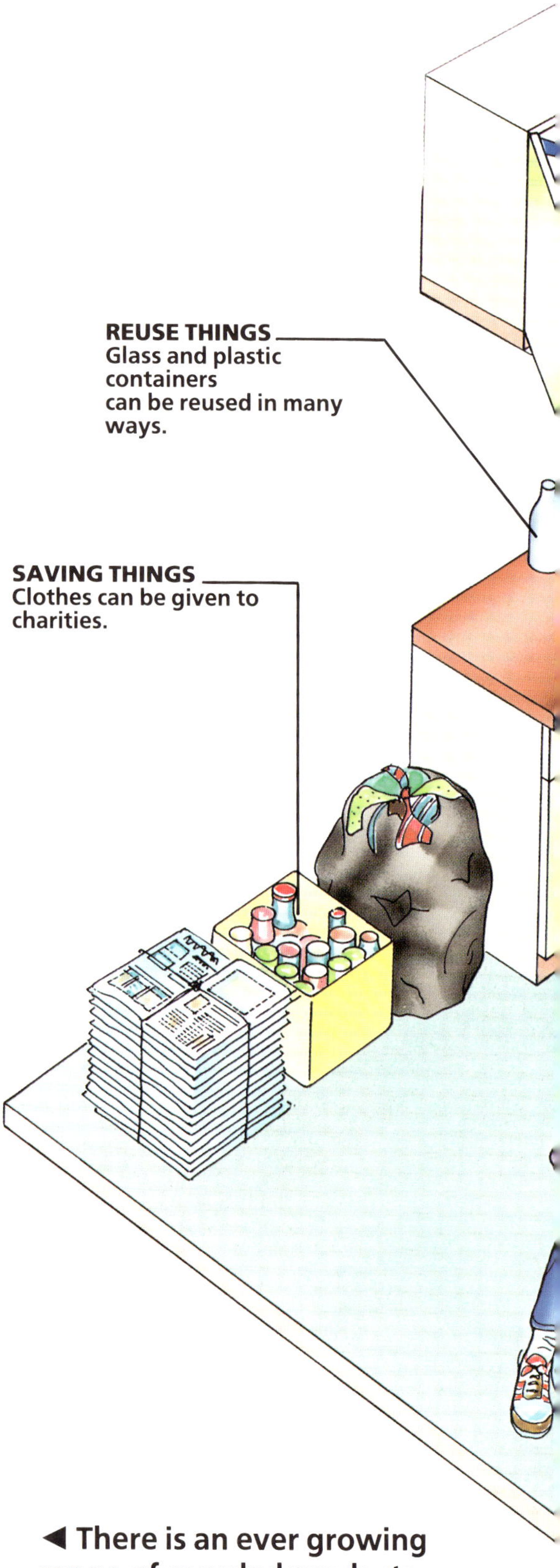

◄ There is an ever growing range of recycled products available. These are often less costly than non-recycled alternatives.

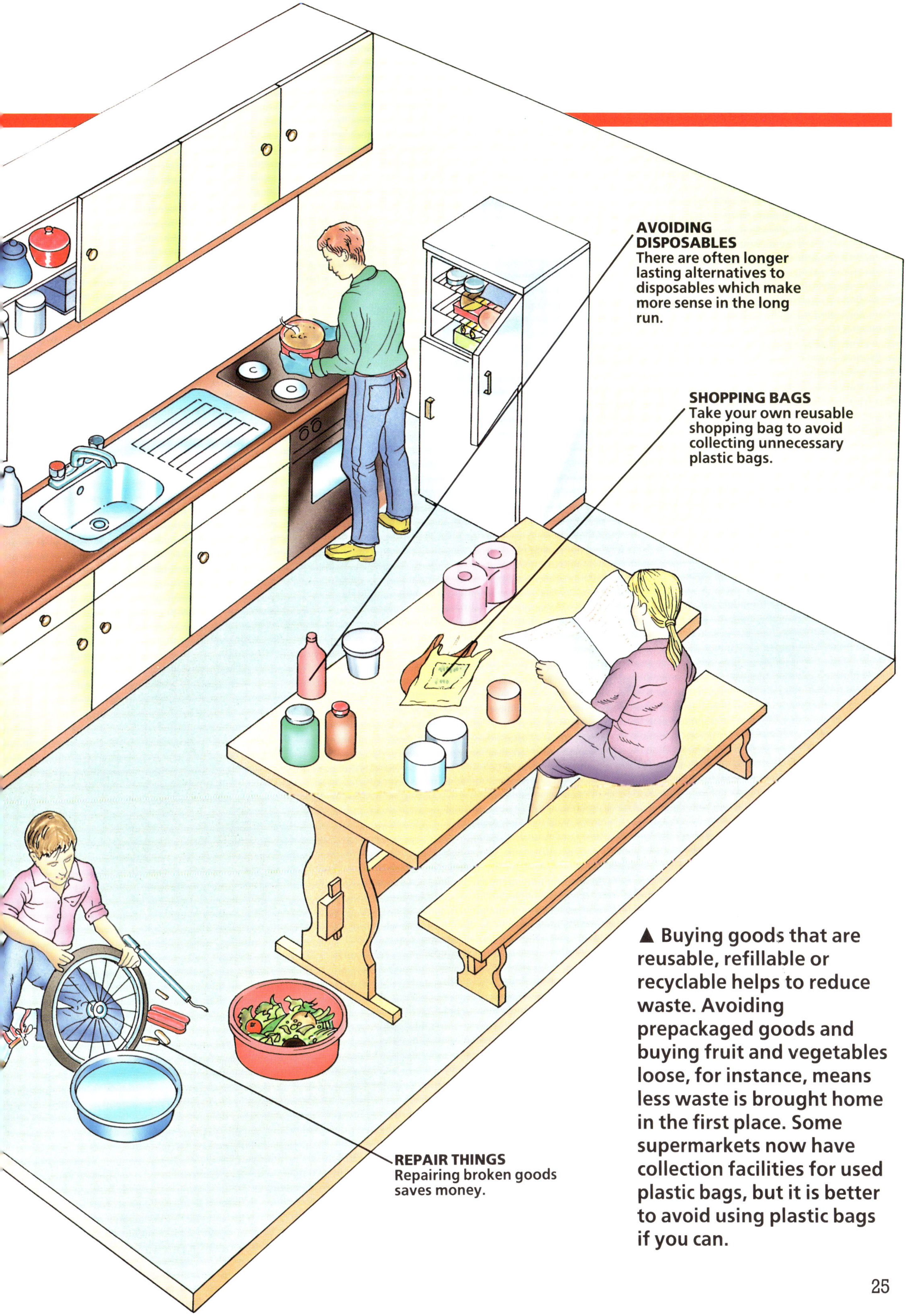

▲ Buying goods that are reusable, refillable or recyclable helps to reduce waste. Avoiding prepackaged goods and buying fruit and vegetables loose, for instance, means less waste is brought home in the first place. Some supermarkets now have collection facilities for used plastic bags, but it is better to avoid using plastic bags if you can.

WHAT YOU CAN DO

You can help to encourage more recycling and reduce waste by:

- Finding out about your local recycling facilities and using them
- Buying goods made from recycled materials
- Reusing and repairing things
- Using refillable containers
- Saving unwanted clothes for charity shops
- Avoiding disposable products and unnecessary packaging.

Useful addresses:

Friends of the Earth
218 D Street SE
Wasking, D.C. 20005

Friends of the Earth has leaflets and other information about recycling.

Waste Watch
P.O. Box 298
Livingston
Kentucky 40445

Waste Watch provides information on all aspects of recycling and local recycling facilities.

DESIGNING A POSTER:

One of the most important things you can do is to make people more aware about the importance of recycling. One way you can do this is to make a poster to display at your school.

1) Think of an eye-catching heading for the poster that will actually make people stop and read it.

2) Design an illustration or symbol like the one shown here, or cut pictures from magazines and make a collage to help convey your message.

3) Read through the book and summarize in about 30-40 words what recycling is all about and why it is important that everyone should recycle things.

4) Again by reading through the book, make some suggestions as to ways in which people can be encouraged to recycle, repair, and reuse more.

5) If you have room, include other information such as useful addresses to contact for more information about recycling, or suggestions of your own on reducing waste.

RECYCLE
RE USE · REPAIR
IN OUR CONSUMER SOCIETY THE CONSTANT PRODUCTION OF GOODS LEADS TO A DANGEROUS DEPLETION OF RESOURCES, SOME OF WHICH CANNOT BE REPLACED. THIS ALSO CAUSES WORLDWIDE POLLUTION AND EVERGROWING MOUNTAINS OF WASTE WHICH ARE DIFFICULT TO DISPOSE OF SAFELY.
WHAT CAN YOU DO?
• AVOID GOODS WITH EXCESS PACKAGING.
• COLLECT RECYCLEABLE MATERIALS LIKE PAPER, GLASS AND TIN CANS.
• BUY GOODS BUILT TO LAST A LONG TIME.
• MEND BROKEN ITEMS IF POSSIBLE INSTEAD OF BUYING NEW ONES.
USEFUL ADDRESSES

FACT FILE 1

Recyclable cars
Volkswagen and BMW have experimented to produce cars that are built so that they can be taken apart for recycling at the end of their useful lives. Volkswagen in Germany gives a full take-back guarantee for its new Golf, and is introducing licensed scrapyards that will strip and return recyclable materials to the Volkswagen factories, so that they can be recycled into new car parts. Other car manufacturers are likely to follow suit.

The Tire Mountain
Millions of car and truck tires in the European Community reach the end of their useful life in a year. Nearly half of these will be dumped or left lying around. Tires can be retreaded and used again. The rubber can also be broken down into crumbs and used to make new products or burned to provide energy. Steel and zinc from tires can also be recycled. Tire dumps are often set alight producing poisonous liquids and gases.

COUNTRY	GLASS	ALUMINUM CANS	STEEL CANS
France	34%	negligible	24%
Italy	40%	29%	n/a
Netherlands	53%	negligible	45%
Norway	6%	80%	n/a
Sweden	22%	82%	n/a
Switzerland	53%	26%	n/a
UK	20.4%	9.5%	9.5%
West Germany	39%	negligible	45%

Current Recycling Rates

Even in the countries that are doing most to increase recycling, rates are still relatively low and more needs to be done to encourage people to make use of collection systems.

Disposable Diapers

Disposable diapers make life easier for many parents, but they add considerably to the amount of waste we throw away, and because they contain untreated human sewage they add to the potential health risks from garbage dumps. Each year 3.5 billion diapers are thrown away in the United Kingdom – about 80 tons an hour. There are now more easy-to-use washable diapers available and in the United States most cities have diaper washing services.

FACT FILE 2

Paper Waste from Offices

It is estimated that every office worker in the United Kingdom throws out about 5.3lb of paper waste each week. Paper makes up 50-80% of office waste. Most paper used by offices is of good quality and in demand for making recycled paper for printing and writing, and for things like tissues. It is fairly simple to set up a system to collect waste paper from offices, and companies have found this can save money on disposal costs. Good quality recycled paper has reached a standard high enough to replace the high-grade paper made from woodpulp generally used in offices. Some companies are already looking at using recycled paper. The more companies that do this the more beneficial it will be to the environment.

Labeling Goods

In order to help recycling succeed, better systems of labeling products need to be introduced so that we know what we are buying and what it is made from. It would help if drink cans were labeled to show whether they are made from steel or aluminum, so that there is no need for them to be tested with a magnet. Plastic bottles should be labeled to show what type of plastic they are made from in order to make sorting easier. Recycled paper products should show what type and percentage of recycled paper they contain. There are already in existence environmental labeling systems like the Blue Angel and Green Dot, but a universally agreed system used by everyone would make things easier.

GLOSSARY

Acid rain – rain that is made acidic when pollution from industrial areas and cars reacts with water in the atmosphere. It can damage trees and soil and can acidify lakes. Acid rain is also harmful to health.

Biogas – rotting organic waste in an airtight container gives off a mixture of carbon dioxide and methane which can be burned to provide electricity or heating.

Compost – partly decayed plant material used as a natural fertilizer. It is made up of grass cuttings and plant material from the backyard as well as organic houshold waste. It is used to provide nutrients for the soil and to loosen the soil's structure. Compost should be allowed to decay for five to seven months before it is used.

Dioxins – extremely poisonous chemicals which contain chlorine. They are created during the manufacture of paper and chemicals, and are also produced when plastics are burned at low temperatures.

Fertilizers – substances containing chemicals necessary for healthy plant growth. They are used to compensate for poor soil or soil depleted by repeated cropping.

Greenhouse effect – some gases in the atmosphere let the sun's rays through to warm the earth, but do not allow the radiation coming back to pass through. This keeps the atmosphere – and the planet – warm. It is known as the greenhouse effect and helps make life on Earth possible.

Landfill – an area of land where garbage is dumped, packed down, and covered with soil. Most domestic garbage and industrial waste is dumped in landfill sites.

Mulch – material that is spread over soil to prevent water in the soil from evaporating, but allowing air still to get through.

Organic garden – land that has had no artificial fertilizers or pesticides applied to it for at least two years.

Recycling – reclaiming the useful materials from waste so that they can be used again, such as glass, metals, paper, cloth, plastics, and rubber.

Waste – anything discarded as garbage because it is not needed or wanted. It includes dirty liquids from cleaning processes, and gases produced by burning fuels in engines, power stations and industries.

INDEX

A
acid rain 10, 20, 31
air pollution 9, 10, 13, 28
aluminum 8, 11, 16, 30

B
bauxite 11
biogas 18, 23, 31
bottle banks 14

C
carbon dioxide 6, 7, 11, 31
cars 28
clothing 15, 24
composting 18-19, 31
conifers 10
consumer goods 5

D
dioxins 22, 31
disposable diapers 12, 29
disposable products 12, 25, 29

E
electricity 22, 23
energy 5, 8, 10, 14, 22-3

F
fertilizers 18, 19, 31
finite resources 8, 9, 21
fossil fuels 6, 9

G
garbage dumps 12, 13
glass 6, 8, 15, 16, 22
glass cullet 6, 8
global warming 6
greenhouse effect 6, 20, 31

H
health hazards 12, 29
herbicides 10
household waste 5, 12, 14-15, 22

I
incineration 10, 12, 13, 20, 22
industrial waste 5, 20-1
information sources 26
iron 11

L
labeling products 30
landfill sites 10, 12, 23, 31

M
metals 6, 7, 9, 11, 15, 16, 22, 28, 30
methane gas 12, 23, 31
minerals 9
mulch 18, 31

N
natural resources 5, 8, 9, 10, 21

O
oil 8, 21, 22
organic gardening 18, 31
organic waste 6, 7, 15, 18-19, 23

P
packaging 24, 25
paper production 8, 10, 16, 30
paper waste 15, 16, 30
peat 18
pesticides 10, 18
plastic bags 25
plastics 8, 15, 16, 20, 21, 30

R
raw materials 8, 10, 14
recycled products 8, 16, 24
recycling centers 14, 17
recycling process 6, 16-17, 31
recycling programs 5, 6, 14
recycling rates 16, 29
reducing waste 24-5, 26
resource recovery programs 22
rubber 28

S
soil pollution 9, 10, 12, 19, 22
steel 6, 11, 16, 28, 30

T
tin 16
tires 28

W
waste 5, 7, 12, 14-15, 18-21, 22, 23, 31
waste disposal 10, 12-13, 20
waste exchange programs 20
water pollution 9, 10, 19

Photographic Credits:
Cover and pages 6, 7 bottom, 12 right, 16, 18, 24, 28 bottom and 29: Roger Vlitos; pages 4-5, 8, 13 and 14: Eye Ubiquitous; pages 7 top and 19: Panos Pictures; pages 9 top, 10-11, 11 right and 23 left: Science Photo Library; page 9 bottom: Spectrum Colour Library; pages 11 left and 23 right: The Hutchison Library; pages 12 left and 30: Natural History Photographic Agency; pages 20 and 22-23: Frank Spooner Pictures; page 28 top: The J. Allan Cash Photo Library.

PRINTED IN BELGIUM BY proost INTERNATIONAL BOOK PRODUCTION